ENGINEERING JOB ANALYSIS, DESCRIPTION & SPECIFICATION I

HENRY JOHN N. NUEVA

First Edition

2014

Published by the Author through Lulu.com a Self Publishing Company based in Raleigh, N.C. USA

© 2014 Henry John N. Nueva. All rights reserved.
ISBN – 978-1-304-86564-9

Any copy of this book without the permission from the author or publisher shall be considered as proceeding from an illegal source.

All rights reserved. No part of this book may be reproduced in any form or by means electronic or mechanical, including photography or recording system, without written permission in writing from the author or the publisher.

To

Engr. Alfredo D. Apolinar

Dr. Meliton G. Dassun

Dr. Josefina B. Bitonio

Dr. Reynaldo T. Gelido

a fervent and zealous exponent of enlightened Workforce Management Administration and Engineering Enterprise

As well as to my dearest family Nagun-Nueva

this book is sincerely dedicated...

Preface

Recent development in the field of Engineering and Management is swiftly upward- this is obvious especially in the administering workforce development and transition. The need for more resourceful, economical and equitable management of workforce system in commerce and industry has never been as pronounced as it is today.

Scientific management and methodologies continuous searching for ways to improve the production process and human capital towards better workforce management command yet approaches to the solution of managerial problems in business and engineering enterprise is undeniably developed and recognized. Scientific management application can be found in numerous agencies and companies; when a certain representative or official aims to discover what particular management procedure is applicable in a given situation by carefully examining the job, determining what is to be accomplished and then creating or designing the tools and methods to accomplish the task.

This book includes definitions and a brief coverage of job analysis fundamentals and applications. It is intended to show the practical importance of the material covered towards engineering enterprise' strategic human resource management with emphasis on job analysis methods and job description. Review questions and short cases were also provided at the end to facilitate the ability to install job analysis program.

Contents

Title   *Page*

1. Introduction to Engineering Job Analysis………………………..2

2. Job Analysis: Defined………………………………..................5

3. Job Analysis in Engineering Methodologies ………………………7

4. Job Analyst and Its Functions in Engineering Enterprise…………11

5. Methods of Collecting Information………………………….......13

6. Types of Job Analysis Interview Method…………………………..18

7. The Questionnaire Method & Classifications…………………….....21

8. Job Analysis Control…………………….......................................28

9. Job Description & Specification…………………………………....32

10. Job Specification Information…………………………………….....41

Engineering Job Analysis

INTRODUCTION

Why learn Job Analysis?

Hiring is of key importance to the overall productivity of Engineering Enterprise or any other businesses in private or public settings. A well put together job description is a good business investment since it can be used to support most HR functions like recruitment, selection, orientation, training, work plans, compensation, performance reviews and legal defense. Job descriptions explain the key responsibilities of the actual position, reporting relationships and work environment.

Here are some major reasons why Job Analaysis is important in the development and management of Human Resource Unit:

(1) To comprehend the essence of studying jobs and knowing what each worker does, how he does it, under what conditions he executes his job, and what other requirements each worker must hold to perform his job satisfactorily.

(2) To be able understand the importance of knowing the duties, responsibilities and requirements of each job as a tool in employee selection and hiring and in appraising the employee performance in the job and many other applications.

(3) To learn the methods, mechanics, and guidelines of analyzing the different jobs in the organization and be able to write the job descriptions and job specifications.

(4) To utilize the resources in developing and installing job analysis program in a company or organization.

On the other hand, the purpose of getting greater job satisfaction, increased performance, reduced absenteeism & turnover, and greater profitability is one of the organizational objective where job analysis and programming can be utilize.

Furthermore, Job Analysis as a basic instrument of human resource management in the field of engineering and technology applied more specifically on the following ideal key areas:

Figure 1. Basic Rationale of Job Analysis

Organization and Manpower Planning: It is helpful in organizational planning for it defines labor needs in clear terms. It coordinates the activities of the work force and facilitates the division of work, duties and responsibilities;

Recruitment and Selection: Job analysis indicates the specific job requirements of each job (i.e. skills and knowledge). Through this, JA provides a realistic basis for hiring, training, placement, transfer and promotion of personnel. Basically, one of the goals of the analysis is to match the job requirements with a worker's aptitude, abilities and interests.

Training & Development: Job analysis determines the level of standards of job performance. Job analysis provides the necessary information to the management of training and development programs. It helps to determine the content and subject matter of training courses and helps in checking application information, interviewing, and weighing test results and in checking references.

Wage and Salary Administration: JA is the foundation for job evaluation. It could help by indicating the qualifications required for doing a specified job and the risks and hazards involved in its performance.

Performance Appraisal: It will help in establishing clear cut standards which may be compared with the actual contribution of each individual. Job analysis data provide a clear cut performance for every job.

Job Re-engineering: JA provides information which enables the management to change jobs in order to permit their being manner by personnel with specific characteristics and qualifications that can be either *Industrial Engineering Activity* (IEA) which industrial engineers may use job analysis information in designing the job by making comprehensive study or on the other hand, applying *Human Engineering Activity* (HEA) where study on time, motion and work measurement (i.e. Physical, Mental and Psychological) are studied.

Health & Safety: Job analysis provides an opportunity for identifying hazardous and unhealthy conditions so that corrective measures may be taken to minimize the possibility of accidents and sickness.

What is Job Analysis?

Models of Job Analysis were primarily conducted for the purpose of improving the efficiency of employees. Father of Scientific Management, Frederick Taylor (1911) studied work by breaking it down into smallest identifiable components to find out the best way to perform each component and then compiled work into larger duties that finally lead to jobs.

Job Analysis: Defined

Job Analysis (JA) is the process of studying jobs to gather, analyze, synthesize and report information about job responsibilities and requirements and the conditions under which work is performed (Heneman & Judge, 2009).

Organizations and Engineering Enterprises exist to accomplish some goal or objective. They are collectivities rather than individuals because achieving the goals requires the effort (work) of a number of people (workers). The point at which the work and the worker come together is called a job – it is the role played by worker. To some extent we need to know lot of information about these roles/jobs that includes:

- What does or should the person do?
- What knowledge, skill, and abilities does it take to perform this job?
- What is the result of the person performing the job?
- How does this job fit in with other jobs in the organization?
- What is the job's contribution toward the organization's goals?

The goal of this process is to secure all necessary job data. Job evaluation represents the major use of job analysis. Since job information needed for various uses may differ, some organizations make a specialized study for each specific use.

To identify the jobs in an organization and to differentiate them from one another, each job category must be given a specific title. Other cases however, job titles are too general to suitably describe the duties and responsibilities inbuilt on each. Thus, no standard job descriptions and job specifications can be uniform in all firms. To determine the duties, responsibilities and skills pertinent to each job category, one has to analyze each position in terms of the duties and responsibilities involved, and the relationship of the job to other jobs.

Next to these, one must determine the human traits, skills, knowledge, and abilities of worker to enable him satisfactorily discharge his/her duties and responsibilities inherent in the positions.

Job Analysis: Detailed Uses & Applications

Engineering firms are encouraged to navigate and perform the fundamental application of job analysis. This is to effectively come up with a quality workforce and production output that can be use in measuring level of fineness or make in addition of strategic human resource planning and management.

The primary uses of job analysis (P.S. Sison, 1982) as applied to any organization are explicated as follows:

1.) To know the duties of each job by studying its requirements in terms of skills, efforts, responsibilities and working conditions;
2.) To serve as a guide in the recruitment, selection, placement, and counseling of employees;
3.) To serve as a basis for job evaluation and wage and salary administration;
4.) To develop channels of promotion and transfer along lines determined by the duties, responsibilities, job-requirements, and working conditions;
5.) To help in counseling and handling of grievances;
6.) To help determine working conditions that are hazardous, unpleasant, or unhealthy and thus enable management to institute preventive as well as corrective measures that will improve these conditions;
7.) To serve as a guide in establishing standards of performance, production standards, simplifying work procedures, and improving methods through the analysis of methods and time-and-motion studies;

8.) To help in effective supervision. Supervisors are provided with precise and detailed descriptions of the jobs under their respective jurisdictions and they can make these serve as basis for judging employees performance;

9.) To determine the training needs of an employee who may not yet possess the skills and abilities required by the position he or she occupies;

10.) To standardize job titles that reflect the functions required of each job;

11.) To enable management to institute a job control system in which a continuous inventory of all authorized positions can be made available at all times.

Job Analysis in Engineering Methodologies

What is a Technical Job?

Technical Job skills refer to the talent and expertise a person possesses to perform a certain job or task. This is sometimes called "hard skills", as opposed to soft skills which are personality and character traits. Meanwhile, technical skills are those abilities acquired through learning and practice; in other words a particular skill set or proficiency required to perform a specific job or task.

Approaches to Analyzing Jobs

There is no one way to study jobs even in Engineering Enterprise. Numerous models of job analysis is now exist, each focusing on some particular use for job analysis. This process may seek to obtain technical information about work, workers, and context within which the job exist.

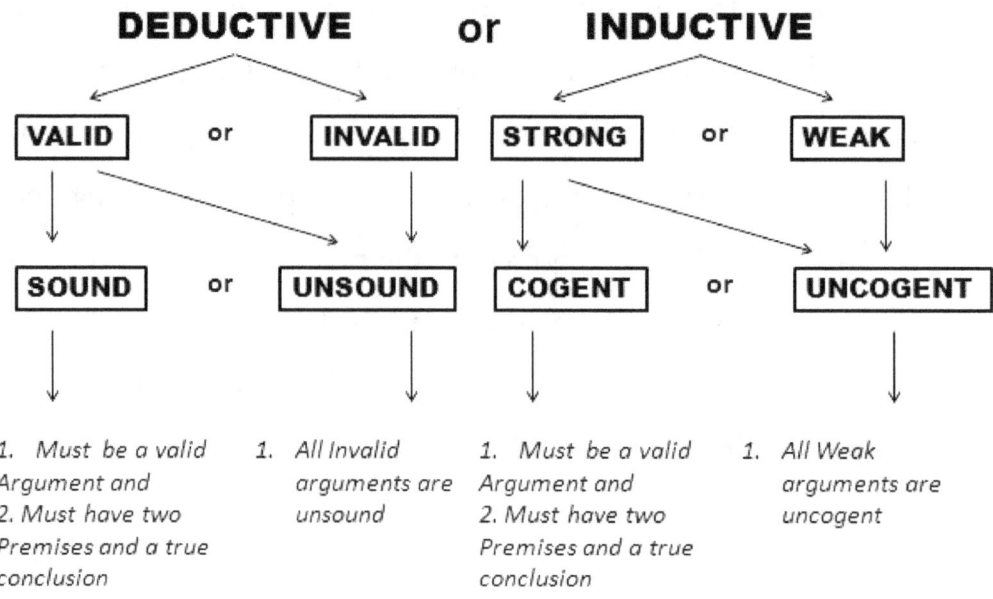

Figure 2. Deductive & Inductive Decision Tree

Further, the approach may be either *inductive* or *deductive* approach. In an inductive approach, information about a job is collected first and then organized into a framework to create a description of a job. While in a deductive approach, a model of information is developed and the collection of data focuses upon this model.

The basic principle of Job Analysis model as formulated in the idea of obtaining information on work activities, a proposed formula can be used in order to facilitate and initiate the functions. This formula includes:

1. What the worker/employee does?
2. How he or she does it?
3. Why he or she does it?
4. Skill involved in doing it.

In fact, providing the WHAT, HOW and WHY of each task and the total job should constitute a functional description of work activities for compensation purposes.

Dimensions of Job Analysis

There is a huge amount of job analysis technique and these techniques may differ on a number of dimensions.

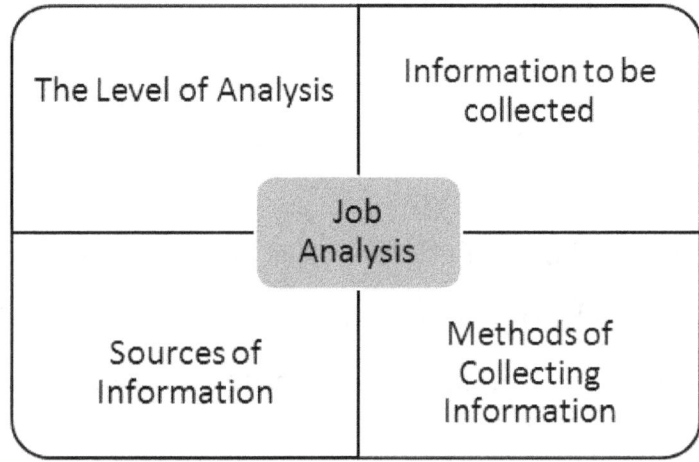

Figure 3. Dimensions of Job Analysis

Level of Analysis

By remembering the concept we're discussing job analysis, we entail that the unit of analysis is the job. In fact, the level or unit of analysis represents a decision that is worthy of argument.

The lowest level is Employee Attributes – the knowledge, skills, and abilities required by the job. One level up is the element. An element is often considered the smallest division of work activity apart from separate motions, although it may be used to describe singular motions. At such, it is the *unit of analysis for time and motion* study, and is used primarily by industrial engineers.

Next level is the Task, a *discrete unit of work* performed by an individual. A task is more independent unit of analysis. It consists of a sequence of activities that completes a work assignment. When sufficient task accumulate to justify the employment of a worker, a positive exists. Furthermore, a job is a group of positions that are identical in their major or significant task. The positions are sufficiently alike to justify being covered by a single analysis and description. One or many persons may be employed on the same job. Jobs found in more than one organization are termed occupations. And therefore, occupations grouped by function are usually referred to as job families.

Information to be collected

Since the job is linked between organization and the employee, it may be useful to develop a model based on this common connection. We can say that both the organization and the employee contribute to the job and expect to receive something from it.

The principal sources of information about a job are the employees assigned to it, other employees who formerly occupied the job and the supervisor of the unit where the job is. The job information to be collected should also include the personal traits required of the worker to perform it excellently.

Below are sample of key points on determining various job-related information cover:

1. *What job requires the worker/employee to do?*
2. *How the workers discharge the duties of the job efficiently?*
3. *Why the work is performed.*
4. *Supervision involved in the job.*
5. *The working conditions.*

The study of jobs is a function of the Human Resource Department (HRD), however today some firms allocate this function to outside consultants (outsource) hired on a contractual basis. This is to update the data and information on jobs and supervisors notify the personnel department about any changes made relative to the duties and responsibilities of the jobs under his authority and directives which is done normally on a periodic scheme.

Job Analyst and its Function in Engineering Enterprise

In today's complex business environment, an organization's adaptability, agility, and ability to manage constant change through innovation can be keys to success. Workforce management traditional methods will no longer lead to reaching objectives under current labor conditions. That's where Job Analysis comes in. Innovative corporations achieve goals through HR Management projects that translate workforce needs into positive deliverables and insights. And Job Analyst makes it all happen more efficiently and effectively.

A *Job Analyst* is a person who studies the duties, responsibilities and other requirements of a job and writes down the corresponding jobs description and specification. Becoming successful Job Analyst takes core business & management skills, and specialized knowledge that will advance a Human Resource Department as well as the Company's objectives and contribute to its remaining competitiveness in a complex administration. These core skills include:

- Written and Verbal communication, including technical writing skills
- Understanding of systems Engineering Concepts
- The ability to conduct time and motion study
- Personnel Management Case Development
- Modeling techniques and methods
- Leadership

Increasingly, modern successful business organization recognizes the value of job analyst training to improve their overall performance. Job Analysts gather information about jobs through interviewing employees, observing performance in certain tasks, asking employees to fill out questionnaires and worksheets, and by collecting further information about the job from other sources. Job Analyst will then writes up their findings from such analysis and review them with management.

After this, the documentation is presented to the senior manager/supervisor for review. He may edit the documentation by adding, deleting or modifying duties, knowledge, skills, abilities, and other characteristics and requirements relevant to the job. A signed and dated job description is then prepared and the job description becomes the official company record for a particular job.

Meanwhile, there are no specific guidelines outlining the requirements needed to become a job analyst in engineering enterprises and companies, however, most have an undergraduate degree on engineering and a masters degree on engineering management or business administration. Some job analysts even have a PhD. The majority of job analysts have extensive education in industrial and organizational psychology. This field of psychology deals with how people perform work tasks and how they interact at the workplace. Such programs focus on statistics and relate to other disciplines such as sociology, economics, and other branches of psychology.

Moreover, below are the duties of a Job Analyst excerpts from the Dictionary of Occupational Titles:

```
CODE: 166.267-018                    Buy the DOT: Download/CD-ROM
TITLE(s): JOB ANALYST (profess. & kin.) alternate titles: personnel analyst

Collects, analyzes, and prepares occupational information to facilitate personnel,
administration, and management functions of organization: Consults with management to
determine type, scope, and purpose of study. Studies current organizational occupational
data and compiles distribution reports, organization and flow charts, and other background
information required for study. Observes jobs and interviews workers and supervisory
personnel to determine job and worker requirements. Analyzes occupational data, such as
physical, mental, and training requirements of jobs and workers and develops written
summaries, such as job descriptions, job specifications, and lines of career movement.
Utilizes developed occupational data to evaluate or improve methods and techniques for
recruiting, selecting, promoting, evaluating, and training workers, and administration of
related personnel programs. May specialize in classifying positions according to regulated
guidelines to meet job classification requirements of civil service system and be known as
Position Classifier (government ser.).
GOE: 11.03.04 STRENGTH: L GED: R5 M4 L5 SVP: 6 DLU: 77
ONET CROSSWALK: 21511A Job and Occupational Analysts
```

Figure 4. Job Analysts Duties and Responsibilities

Methods of Collecting Information

Job Analysis traditionally has been conducted in a number of different ways. Also, firms differ in their needs and in the resources they have for conducting job analysis.

An HR specialist (an HR specialist, job analyst, or consultant), a worker, and the worker's supervisor usually work together in conducting job analysis. Since job analysis data is usually collected from several employees from different department, using interviews and questionnaires. The data is then averaged, taking into account the departmental context of the employee, to determine how much time a typical employee spends on each several specific tasks.

The Interview Method

This method is most commonly used in gathering facts about job performed inside the company. The Job Analyst obtains information about the work by personally conferring with the worker or the latter's supervisor, and sometimes with both, either in the place of work or in the supervisor's office.

There are three (3) basic types of interviews managers use to collect job analysis data namely:

1. *Individual* (this is to get the employee's perspective on the job's duties and responsibilities)
2. *Group* (applies when large numbers of employees perform the same job)
3. *Supervisor* (to get his/her perspective on the job's duties and responsibilities)

A job analyst normally prepares a form to use as his guide in gathering facts about the job. Through the interview method, job analyst acquires a personal impression to the job, and this contributes to the accuracy of his job later.

JOB ANALYSIS INFORMATION FORMAT

Your Job Title _____ Code _____ Date _____

Class Title _____ Department _____

Your Name _____ Facility _____

Supervisor's Title _____ Prepared by _____

Superior's Name _____ Hours Worked _____ AM/PM to _____ AM/PM

1. What is the general purpose of your job?

2. What was your last job? If it was in another organization, please name it.

3. To what job would you normally expect to be promoted?

4. If you regularly supervise others, list them by name and job title.

5. If you supervise others, please check those activities that are part of your supervisory duties:
   - __ Hiring
   - __ Orienting
   - __ Training
   - __ Scheduling
   - __ Developing
   - __ Coaching
   - __ Counseling
   - __ Budgeting
   - __ Directing
   - __ Measuring performance
   - __ Promoting
   - __ Compensating
   - __ Disciplining
   - __ Terminating
   - __ Other _____

6. How would you describe the successful completion and results of your work?

7. Job Duties—Please briefly describe what you do and, if possible, how you do it. Indicate those duties you consider to be most important and/or most difficult.

   a. Daily duties—

   b. Periodic duties (please indicate whether weekly, monthly, quarterly, etc.)—

   c. Duties performed at irregular intervals—

   d. How long have you been performing these duties?

   e. Are you now performing unnecessary duties? If yes, please describe.

   f. Should you be performing duties not now included in your job? If yes, please describe.

Figure 5. An example of (JAIF) Job Analysis Information Form

8. *Education.* Please check the blank that indicates the educational requirements for the job, not your *own* educational background.

   a. _____ No formal education required.

   b. _____ Less than high school diploma.

   c. _____ High school diploma or equivalent.

   d. _____ Two-year college certificate or equivalent.

   e. _____ Four-year college degree.

   f. _____ Education beyond undergraduate degree and/or professional license.

   List advanced degrees or specific professional license or certificate required.

   Please indicate the education you had when you were placed on this job.

9. *Experience.* Please check the amount needed to perform your job.

   a. _____ None.

   b. _____ Less than one month.

   c. _____ One month to less than six months.

   d. _____ Six months to one year.

   e. _____ One to three years.

   f. _____ Three to five years.

   g. _____ Five to 10 years.

   h. _____ Over 10 years.

   Please indicate the experience you had when you were placed on this job.

10. *Skills.* Please list any skills required in the performance of your job. (For example, degree of accuracy, alertness, precision in working with described tools, methods, systems)

    Please list skills you possessed when you were placed on this job.

11. *Equipment.* Does your work require the use of any equipment? Yes _____ No _____ If yes, please list the equipment and check whether you use it rarely, occasionally, or frequently.

Equipment	Rarely	Occasionally	Frequently
a. _____	_____	_____	_____
b. _____	_____	_____	_____
c. _____	_____	_____	_____
d. _____	_____	_____	_____

Figure 5. …continuation of (JAIF) Job Analysis Information Form

Using Interview method in gathering facts for Job Analysis provide numbers of advantages. One of this is the fact that information and question are more comprehensive because it is simple, quick and the interviewer can unearth activities that may never appear in written form.

Often an employee may feel uncomfortable being interviewed for a Job Analysis because they may feel that the results of the job analysis will adversely affect them in terms of salary or working conditions. However, in order to facilitate a successful interview towards job analysis, the following ***Basic Guidelines & Tips*** are genuinely suggested:

- **Be Prepared**
    - *Formulate and practice a list of questions to be asked to the interviewee.*
    - *Consider the background of workers when planning interview questions.*
    - *Anticipate questions about the interview process from the worker.*
    - *Make sure that all forms and documents are present and are easily read and understood.*
- **Conduct the interview in a setting free from noise and other distractions**
- **Attempt to conduct the interview in pairs. It makes recording of information easier**
- **Begin by introducing yourself in an informal manner and by clearly explaining purposes of the interview**
- **Establish rapport**
    - *Give the interviewee an opportunity to get used to the situation and to you. After the introduction, start by making general comments and by asking broad questions.*
- **Keep interview length to under an hour**

- **Record interview information by any technique you feel comfortable with although some type of standard recording is preferred. However, use of tape recorders is not advised since workers feel sometimes uneasy and are time consuming to transcribe**
- **Actively listen to the answers of workers**
    - *Repeat questions and/or give an example in response to the worker's answer.*
    - *Limit the amount of time you record responses of the worker on paper.*
- **Avoid questions that can be answered by yes or no responses**
- **Avoid questions that begin with the word "Why"**
- **Avoid "leading" questions**
- **Ascertain the specific activities performed by the worker**
    - *It is useful to ask the employee to think of a typical work day and tell you the first task they perform when they walk in.*
    - *To achieve the desired detail regarding what the worker does, it is beneficial to focus on the inputs and the behaviors that the workers engages in (outputs) to respond to these inputs.*
    - *Questions of the following type can prove useful in this regard:*
        - ❖ What specific activities are performed when a given input is received?
        - ❖ What are the specific outputs of the work performed?
        - ❖ Where, if anywhere, are these outputs sent?
        - ❖ What, if any, time requirements exist to produce the outputs?
- **Exercise a moderate amount of "control" during the interview**
- **Be aware of the expressions and reactions of the worker**
- Conclude with informal talk and summary of the interview

- ✓ Ask the worker if he/she has any questions
- ✓ Request any additional general information or clarification
- ✓ Reconfirm the purpose and intended use of the interview information
- ✓ Don't forget to give thanks for their time

Types of Job Analysis Interview Method

In preparing job interview questions and scripts, it is essential to distinguish what type of interview you will need to use. Unstructured Interviews are very informal and produce qualitative data (descriptive data). The analyst can ask the participant questions based on their previous answers, giving the participant opportunity to raise related topic and take the interview in their chosen direction.

Unstructured Interview is a conversation with no prepared questions or predetermined line of investigation. However the analyst should explain (1) the purpose of the study and (2) the particular focus of the interview.

Formal	Engineered		Collaborative
Structured		**Unstructured**	
Pushed	Active	Pulled	Unconscious
Planned	Flexible	Unplanned	Unsolicited
Rigid	Conscious	Support	Instinctive
Mandated	Just-in-time	Just-enough	Innate
Passive	On-demand	Self-service	Contextual

Figure 6. Structured & Unstructured Interview in a single case design

The roles and the purposes give framework. The analyst generally uses a questioning strategy to explore the work the job holder performs. Listening and taking notes are very important. These enable follow up questions to be posed. The questions and responses —with summaries enable the interview to be controlled. The conversation takes on a structure with areas being considered, explored, and related to each other and revisited secure in depth of information required during analysis.

An unstructured interview involves *questions and responses* and may be free flowing but it becomes structured in the sense that the analyst has a purpose and needs skill to:

1. *Establish a relationship*
2. *Ask well-structured questions to generate a conversational flow in which the worker offers – information- factual, opinion, subjective and objective about aspects of the job*
3. *To ensure information received is heard and understood - listening, clarifying and reflective summarizing*

Effective listening requires concentration and this can be distributed by interruptions, the analyst's own thought processes and difficulty in remaining neutral about want is being said. Notes need to be taken without loss of good eye contact. Cues also need to be picked up so that further questions can be asked to probe issues and areas of interest.

On the other hand, *Structured Interview* is very formal. The workers will be asked set of questions which have previously been carefully thought out in order to get the best findings for the research. This type of interview may assume definite format involving: (a) charting a job-holder's sequence of activities in performance and (b) an inventory questionnaire may be used. Care is needed to set up such interactions. A specialist analyst is not involved and participants need to know what they are doing why and what is expected as a result. They may be entrained as interviewers and not structure the interview as recommended. Also, notes and records may be needed for subsequent analysis.

Interview Outcomes

Interviewing is a flexible method for all levels and types of job since an interview may focus on what a hypothetical job might involve. This method generates descriptive data and enables job-holders to interpret their activities.

A good interviewer can probe sensitive areas in more depth. Structured questionnaires cannot easily do this. Jobholders or employees can give overviews of their work and offer their perceptions and feelings about their job and the environment. Rigid questionnaires tend to be less effective where the more affective aspects of work are concerned. However, information from different interviews can be:

1. *Hard to bring together*
2. *There is a potential for interviewer bias*
3. *Certain areas of the work may fail to be picked up*
4. *An interview may stress one are and neglect others*
5. *There are problems in interpretation and analysis with the possibility of distorted impressions*
6. *The subjectivity of the data captured needs to be considered*

Interviewing as the sole method of job analysis in any particular project has disadvantages. Interviews are sometimes time consuming and training is needed. Co-counseling may remove the analyst and enable the workers to discus work between themselves. Though experience however they may miss items and there is the natural problem of people not establishing and maintaining rapport with each other during an interview.

The Questionnaire Method & Classifications: CMQ, PAQ, & WPS

Questionnaires may be considered self-administered interviews that are typically very carefully structured and pretested. Often the items on a questionnaire are tasks or activities, and workers are asked to evaluate the task on one or more different scales. One such scale might be how difficult each task is to perform.

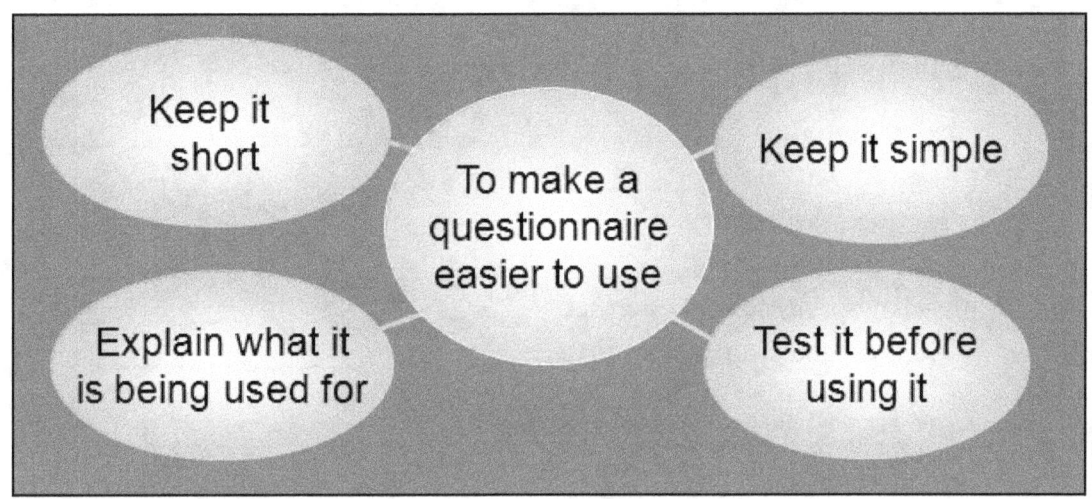

Figure 7. Factors of Good Questionnaire Data

The Common Metric Questionnaire (CMQ) is targeted toward both exempt and nonexempt jobs. This questionnaire classification has five (5) sections namely; (1) Background, (2) Contacts with People, (3) Decision Making, (4) Physical and Mechanical Activities, & (5) Work Setting. The Background section asks 41 general questions about work requirements such as travel, seasonality, and licensure requirements. The Contacts with people section asks 62 questions targeting level of supervision, degree of internal and external contacts, and meeting requirements.

The 80 Decision Making items in the CMQ focus on relevant occupational knowledge and skill, language and sensory requirements, and managerial and business decision making. The Physical and Mechanical section contains 53 items that focus on environmental conditions and other job characteristics. The CMQ is a relatively new instrument. It has been field tested on 4, 552 positions representing over 900 occupations in the Dictionary of Occupational titles (DoT), and yielded reasonably high reliabilities. (Harvey, 1993)

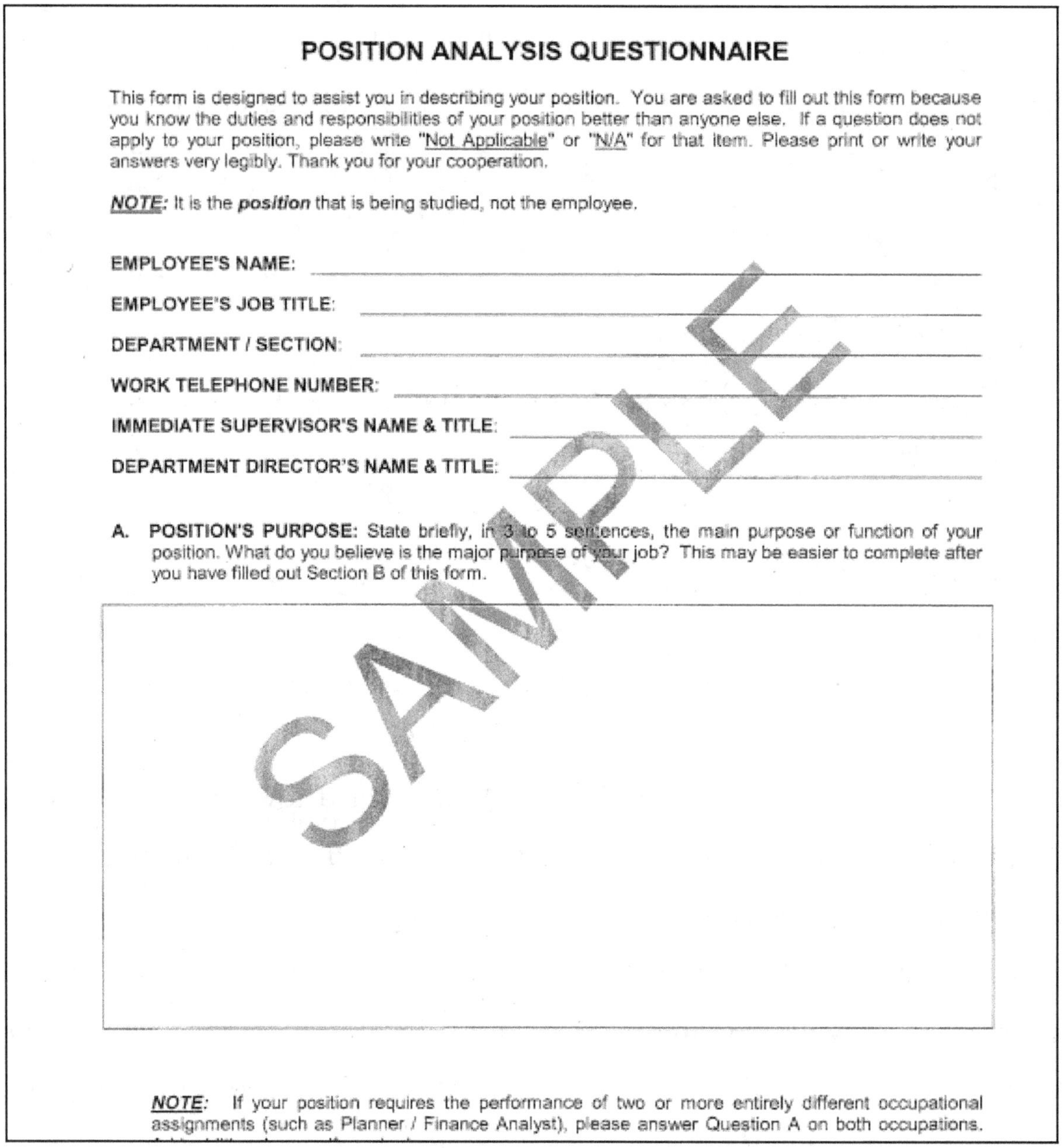

POSITION ANALYSIS QUESTIONNAIRE

This form is designed to assist you in describing your position. You are asked to fill out this form because you know the duties and responsibilities of your position better than anyone else. If a question does not apply to your position, please write "Not Applicable" or "N/A" for that item. Please print or write your answers very legibly. Thank you for your cooperation.

NOTE: It is the ***position*** that is being studied, not the employee.

EMPLOYEE'S NAME: ____________________

EMPLOYEE'S JOB TITLE: ____________________

DEPARTMENT / SECTION: ____________________

WORK TELEPHONE NUMBER: ____________________

IMMEDIATE SUPERVISOR'S NAME & TITLE: ____________________

DEPARTMENT DIRECTOR'S NAME & TITLE: ____________________

A. POSITION'S PURPOSE: State briefly, in 3 to 5 sentences, the main purpose or function of your position. What do you believe is the major purpose of your job? This may be easier to complete after you have filled out Section B of this form.

NOTE: If your position requires the performance of two or more entirely different occupational assignments (such as Planner / Finance Analyst), please answer Question A on both occupations.

Figure 8. Sample PAQ Form

The Position Analysis Questionnaire (PAQ) developed by McCormick, Jeanneret, and Mecham (1972) is a structured job analysis instrument to measure job characteristics and relate them to human characteristics.

In its simplest term, PAQ is a set of questions used to describe the duties and responsibilities of a particular job to determine what a specific job actually entails and what traits are necessary to do it well.

The primary purpose is to compare relationship between aptitude test scores and what skills are actually necessary for good on-the-job performance. It is consists of 195 job elements that represent in a comprehensive manner the domain of human behavior involved in work activities. These items falls under five (5) categories namely:

1. **Information Input** (where & how the worker gets information)
2. **Mental Processes** (reasoning and other processes that workers use)
3. **Work Output** (physical activities and tools used on the job)
4. **Relationships with other persons**
5. **Job Context** (the physical and social context of work)

Preparing a PAQ requires no complex procedure since some basic frameworks are up and ready to follow. Below are the basic technical sections that you may use as a guideline in preparing such PAQ form:

Section I: General Information
- This section shall contain the basic information of the worker that includes:
    - ✓ *Name*
    - ✓ *Current Date*
    - ✓ *Current Job Title (system)*

- *Working Job Title*
- *Name of the Department*
- *Organization*
- *Supervisor's Name*
- *Supervisor's Title*
- *Job Category*
- *Length of time in current position*

Section II: Organizational Relationships

- This section identifies reporting relationships. The purpose is to state the position that provides work direction, complete performance reviews and provides guideline, counseling and/or corrective /disciplinary action relative to worker's position.

- Some of the basic questions that you may include are:
  - *To whom do you report?*
  - *Who reports to you?*
  - *Attach a copy of the current department/unit organizational chart*

Section III: Essential Functions (Duties & Responsibilities)

- Indicate the principle duties and responsibilities performed in this job. Group job duties into major areas and then list the associated tasks in order of importance, beginning with the most important.

- Indicate the approximate percentage of time spent on each task. When indicating time spent, make sure to state the time reference (% of day, month, quarter or year) appropriate to the task.

- Be as descriptive as possible- indicating the desired outcome (or reason) why a specific task is performed.

Section IV: Additional Job Information

- Indicate the specific knowledge, skills, and abilities your job requires. Think in terms of recruiting to fill a job identical to the worker.
- List formal guidelines, regulations and policies to follow and understand to perform a job.

Section V: Decision Making & Problem Solving

- In this section, think about the types of issues a workers position in held accountable/responsible for completing. What decisions need to consult or notify an immediate supervisor or before taking action.
- Types of issues, concerns, and problems come to you to resolve is also necessary.

Section VI: Working Environment

- This area describes the physical conditions in which a job is performed on a regular basis including lifting, pushing, climbing, walking, exposure to different environmental influences as well as estimated percentage of time.
- You may also include the description of mental and emotional environment under which a worker regularly perform the job
- The type of equipment, tools, instruments, machine or other similar objects used are also important to list.

Section VII: Minimum Qualifications

- This section will check the qualification of such worker. Certifications, licenses or registrations require.
- Indicate also the minimum level of education that would prepare someone to perform his/her job

- Minimum number of years of prior experience as well as the type of experience needed to be prepared to perform the job

Section VIII: Primary Purpose (Position Objective)
- This is where a worker can tag or lists down a summary of his/her job.

Another important classification of Job Analysis Questionnaire is the Saville & Holdsworth's, Ltd. –SHL Group *Work Profiling System* (WPS) designed to help employee accomplish human resource functions. It is design to yield reports targeted toward various human resource functions such as individual development planning, employee selection, and job description.

There are three (3) versions of WPS tied to types of occupations namely: managerial, service, and technical occupations. The WPS is computer-administered on-site at a company. It is normally contains structured questionnaire which measures ability and personality attributes in areas such as:

1. Hearing Skills
2. Sight
3. Taste
4. Smell
5. Touch
6. Body Coordination
7. Verbal Skills
8. Number Skills
9. Complex Management Skills
10. Personality
11. Team Role

Some of the basic techniques and purposes of the WPS in determining quality questionnaire methods of Job Analysis are listed below:

Choose defensible assessment procedures and methods
<ul><li>Identify jobs for rotation and cross training</li><li>Identify and define competencies of hiring purposes, career, development and succession planning</li><li>Document existing jobs in the form of job descriptions and person specification</li><li>Structure and define new jobs</li><li>Identify performance review measures</li><li>Create job families</li><li>"Transport" validation research</li><li>Create an equitable pay structure</li><li>Identify specific areas of training & development</li><li>Audit skills requirement</li><li>Develop performance management system</li><li>Re-define and merge jobs</li><li>Match candidates against job profiles</li><li>Improve interviewing</li><li>Build Teams</li><li>Manage changes in roles and culture</li></ul>

After determining the purpose and techniques in developing WPS method, information is collected about a job in structured way from 2 to 4 people who are currently doing the job through questionnaires then info. Will computerize to analyze and meet a variety of objectives.

Job Analysis Control: Request, Review and other Considerations

In every organization of any size, the organization structure, work assignments, job duties and responsibilities are subject to change. New job are established and existing jobs may change either abruptly as a result of re-organization or gradually over time.

In such cases, it is necessary to analyze and describe new or revised jobs in order to assure the proper evaluation. As part of controlling the process of job analysis interview, it is then requested to the management through a supervisors' initiation and reviewed by the Department Head; when a new job has been established under the approved organization structure or the duties of an existing job have significantly changed, the supervisor concerned will initiate a "Request for Job Analysis" form and submit it through proper channels to the Department Head concerned. In his request, the job supervisor will outline the main duties of the new job or the principal revisions.

It is then the Department Head concerned will review the request, ensure its validity, and forward it to the Human Resources Department for necessary action. Upon receiving an approval request for Job Analysis, the HR Department will ensure that the job conforms to the approved organization chart, conduct the necessary job analysis interview and compile the information on the Job Analysis Form.

In carrying out the analysis of a job and completing the required for, the following basic aspects must be considered:

1. *The facts obtained and recorded must refer to the job and not the job incumbent.*
2. *The duties and responsibilities must be for the job, as it exists at present rather than for what the job should be or is thought to be. The minimum requirements must be adequate to support the satisfactory performance of such duties and responsibilities.*
3. *The job facts must, in all cases, be verified to ensure that they are accurate, factual and realistic.*
4. *The duties of each job must be coordinated with the duties of other jobs, above and below, in the organizational unit and with related jobs in other parts of the organization.*
5. *Each duty must be analyzed to ensure that is essential to the operation of the unit.*
6. *Jobs which are similar in nature and of an equal level of difficulty should be combined, wherever possible, under the same job title to provide uniformity whilst still permitting flexibility in work assignment.*
7. *Where there is more than one incumbent in a job, only one job analysis interview is required.*
8. *Where the job has more than one application, such as Secretary or Clerk, it is advisable to obtain details of the applications in the various units in order to ensure adequate coverage.*

REVIEW QUESTIONS:

1. Create a comprehensive characterization of Job Analysis and demonstrate.
2. What are the principal uses or applications of Job Analysis in the Field of Engineering and its manpower?
3. How one does establish the duties, skills, and responsibilities on each engineering job categories (example: Industrial, Chemical, Manufacturing, etc.)
4. Why is Job Analysis a continuing program in the field of Business & Engineering?
5. What are the essential characteristics or qualifications of a Job Analyst?

6. What method/s of Job Analysis do you consider the most useful in determining concise and valid information?

7. What are the methods used in Job Analysis. Provide the advantages & disadvantages of each.

8. Why do you think Job Analysis is an essential tool in developing Human Resource Strategy in implementing quality service in an Engineering Enterprise as well as both private and public companies? Use the space given below.

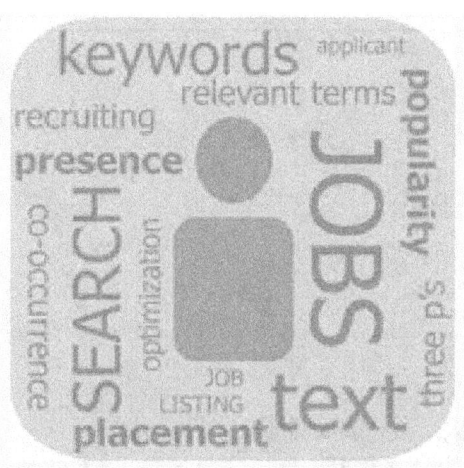

The Job Description & Specification

What is a Job Description & Specification?

Job Description (JD), as a management tool, can greatly simplify an organization's human resource management. A job description clarifies work functions and reporting relationship, helping workers to understand their specific job. Job descriptions aid in maintaining a consistent salary structure and somehow performance evaluation may be based on JD.

```
                                                    Class Code 8031

                        CITY OF CENTRALIA
                    CLASSIFICATION DESCRIPTION

    JOB TITLE:      CIVIL ENGINEER        REPORTS TO:   CITY ENGINEER

    DEPARTMENT:     ENGINEERING           DATE:         November 2006

    GENERAL FUNCTION
    Performs professional engineering work and/or administration and consulting work in the design,
    investigation, and construction of street, utility, ground water, domestic water, storm water,
    wastewater, light systems and other public works projects. Prepares plans and specifications; and
    acts as a project manager on assigned projects.

    This classification description reflects the general concept and intent of the classification and
    should not be construed as a detailed statement of all the work requirements that may be inherent
    in a position.

    JOB DUTIES AND RESPONSIBILITIES

    Essential Functions
    – Duties may include but are not limited to the following:

        •Prepare engineering designs, specifications, costs and quantity estimates of public work
         projects; obtain required easements or permits for streets and light department, water and
         sewer system construction, water, storm water and wastewater treatment facilities, and
         utility structures.

        •Provide engineering support to development review process, including support during
         hearings, pre-application conferences and planning commission meetings.

        •Prepare requests for proposals and bids; review contract bids and proposals; assist in the
         coordination and review of consultants' design work.

        •Review and check adequacy and accuracy of computations, preliminary layout and design
         work from the field and survey data.
```

Figure 9. A Sample of Engineering Job Description

In simplest term, *Job Description* may be defined as an abstract of information derived from the job analysis report, describing the duties performed, the skills, training, and experience required, the conditions under which the job to the other jobs in the organization.

Meanwhile, *Job Specification* (Job Specs.) is the statement of the qualifications and traits required of the worker so that the/she may perform the job properly. It specifies the type of employee which the job calls for in terms of skills, experience, training, and other special qualifications (P. Sison, 1982).

Well written duty statements contain action words which accurately describe what is being done. Duty statement should focus on primary, current, normal, daily duties and responsibilities of such position (not incidental duties and employee's qualifications or performance, or temporary assignments). Related or similar duties should be combined and written as one statement.

Each duty statement should be a discreet, identifiable aspect of the work assignment, described in one to three sentences, and should be outcome-based, allowing for alternate means of performing the duty, changes in technology, preferences of workers and supervisors, and accommodations of workers with disabilities, without altering the nature of the duty itself.

A Duty Statements normally contains three (3) parts: the Verb, the Object, and the Purpose. Take a look at the table below for examples of functional duty statements:

Verb	Object	Purpose
Collects	Financial data	To evaluate budgetary request
Conducts	Analytical studies	To support financial planning & management
Compiles	Enrollment data	For distribution to administration
Cleans	Computer equipment	In adherence with established schedule

Drives	Pickup truck with motor fuels	To job sites
Repair	Equipment	Daily, or as needed

Table 1. A Sample of a good Duty Statement

Considering a good planning and design for job description, the study of any job shall consist of four basic parts namely:

1. **Functions description**
   - This part should identify the description of the functions including duties performed and the responsibilities involved and also the relation of the job with other jobs in the company.

2. **Minimum personal qualification**
   - This section identifies the specification of the minimum personal qualifications n terms of trait, skill, knowledge and ability required of a worker to discharge his job.

3. **Verification & confirmation**
   - On this stage, Draft must be submitted to the supervisor concerned for verification and confirmation.

4. **Job identification**
   - This part will identify the final job by its correct title. And therefore, the title must be reflective of the functions or duties of the job.

On the other hand, an excellent job description basic guideline should identify and describe the following:

MENTAL FUNCTIONS	OPERATIONAL DESCRIPTION
1. Comparing	Judging the readily observable functional, structural or

		compositional characteristic of data, people or things.
	2. Copying	Transcribing, entering or posting data
	3. Computing	Performing arithmetic operations and reporting on and/or carrying out a prescribed action in relation to them.
	4. Compiling	Gathering, collating, or classifying information about data, people or things. Reporting and/or carrying out a prescribed action in relation to the evaluation is frequently involved.
	5. Analyzing	Examining & evaluating data. Presenting alternative actions in relation to the evaluation is frequently involved.
	6. Coordinating	Determining time, place and sequence of operations or action to be taken on the basis of analysis of data.
	7. Synthesizing	To combine or integrate data to discover facts and/or develop knowledge or creative concepts and/or interpretations.
RELATIONS WITH OTHERS		**OPERATIONAL DESCRIPTION**
	1. Supervision (given)	Coordinating and directing the activities of one or more subordinates
	2. Supervision (received)	Independence of actions; authority to determine methods of operation
	3. Negotiating	Exchanging ideas, information and opinions, with others to formulate policies and programs and/or jointly arrive at decisions, conclusions, solutions or solve disputes
	4. Communicating	Talking with and/or listening to and/or signaling people to convey or exchange information that includes giving/receiving assignments and/or directions
	5. Instructing	Teaching subject matter to others, or training other through explanation, demonstration and supervised practice; or making recommendations on the basis of technical disciplines.
	6. Interpersonal Skills and Behavior	Dealing with individuals with a range of moods and behaviors in a tactful, congenial, personal manner so as not alienate or antagonized them.
	7. Control of Others	Seizing, holding, controlling, and/or otherwise subduing violent, assaultive, or physically threatening persons to defend oneself or

PHYSICAL DEMANDS (Strengths)	OPERATIONAL DESCRIPTION
	prevent injury.
1. *Sedentary*	Exerts up to 10 lbs. of force occasionally an/or negligible amount of force frequently or constantly to lift, carry, push pull, or otherwise move objects, including the human body. involves sitting most of the time, but may involve walking or standing for brief periods of time.
2. *Light*	Exert up to 20 lbs. of force occasionally, and/or up to 10 lbs. of force frequently, and/or a negligible amount of force constantly to move objects. Physical demands are in excess of those of Sedentary work. Light work usually requires walking or standing to a significant degree.
3. *Medium*	Exert up to 50 lbs. of force occasionally, and/or up to 20 lbs. of force frequently, and/or up to 10 lbs. of force constantly to move objects.
4. *Heavy*	Exert up to 100 lbs. of force occasionally, and/or up to 50 lbs. of force frequently, and/or up to 20 lbs. of force constantly to move objects.
5. *Very Heavy*	Exert in excess of 100 lbs. of force occasionally, and/or in excess of 50 lbs. of force frequently, and/or in excess of 20 lbs. of force constantly to move objects.

PHYSICAL DEMANDS (Movement)	OPERATIONAL DESCRIPTION
1. *Climbing*	Ascending or descending using feet and legs and/or hands and arms. Body agility is emphasized.
2. *Balancing*	Maintaining body equilibrium to prevent falling on narrow, slippery, or erratically moving surfaces; or maintaining body equilibrium when performing feats of agility.
3. *Stooping*	Bending body downward and forward. This factor is important if it occurs to a considerable degree and requires full use of the lower extremities and back muscles.

4. Kneeling	Bending legs at knees to come to rest on knee or knees.
5. Crouching	Bending body downward and forward by bending legs and spine.
6. Crawling	Moving about on hands and knees or hands and feet.
7. Reaching	Extending hand(s) and arm(s) in any direction.
8. Handling	Seizing, holding, grasping, turning, or otherwise working with hand or hands. Fingers are involved only to the extent that they are an extension of the hand.
9. Fingering	Picking, pinching, or otherwise working primarily with fingers rather than with the whole hand or arm as in handling.
10. Feeling	Perceiving attributes of objects, such as size, shape, temperature, or texture, by touching with skin, particularly that of fingertips.
PHYSICAL DEMANDS (Auditory)	**OPERATIONAL DESCRIPTION**
1. Talking	Expressing or exchanging ideas by means of the spoken word. Talking is important for those activities in which workers must impart oral information to clients or to the public, and in those activities in which they must convey detailed or important spoken instructions to other workers accurately, loudly, or quickly.
2. Hearing	Perceiving the nature of sounds. Used for those activities which require ability to receive detailed information through oral communication, and to make fine discriminations in sounds, such as when making fine adjustments on running engines.
PHYSICAL DEMANDS (Taste/Smell)	**OPERATIONAL DESCRIPTION**
1. Taste & Smell	Distinguishing, with a degree of accuracy, differences or similarities in intensity or quality of flavors and/or odors, or recognizing particular flavors and/or odors, using tongue and/or nose.
PHYSICAL DEMANDS (Vision)	**OPERATIONAL DESCRIPTION**
1. Near Acuity	Clarity of vision at 20 inches or less. Use this factor when special and minute accuracy is demanded.
2. Far Acuity	Clarity of vision at 20 feet or more. Use this factor when visual

		efficiency in terms of far acuity is required in day and night/dark conditions.
3.	*Depth Perception*	Three-dimensional vision. Ability to judge distances and spatial relationships so as to see objects where and as they actually are.
4.	*Accommodation*	Adjustment of lens of eye to bring an object into sharp focus. Use this factor when requiring near point work at varying distances.
5.	*Color Vision*	Ability to identify and distinguish colors.
6.	*Field of Vision*	Observing an area that can be seen up and down or to right or left while eyes are fixed on a given point. Use this factor when job performance re-quires seeing a large area while keeping the eyes fixed.
ENVIRONMENTAL CONDITIONS & PHYSICAL SORROUNDINGS		**OPERATIONAL DESCRIPTION**
1.	*Weather Exposure*	Exposure to hot, cold, wet, humid, or windy conditions caused by the weather.
2.	*Extreme Cold*	Exposure to non weather-related cold temperatures.
3.	*Extreme Heat*	Exposure to non weather-related hot temperatures.
4.	*Wet and/or Humid*	Contact with water or other liquids; or exposure to non weather-related humid conditions.
5.	*Noise*	Exposure to constant or intermittent sounds or a pitch or level sufficient to cause marked distraction or possible hearing loss.
6.	*Vibration*	Exposure to a shaking object or surface. This factor is rated important when vibration causes a strain on the body or extremities.
7.	*Atmospheric Conditions*	Exposure to conditions such as fumes, noxious odors, dusts, mists, gases, and poor ventilation that affect the respiratory system, eyes or, the skin.
8.	*Confined/ Restricted Working Environment*	Work is performed in a closed or locked facility providing safety and security for clients, inmates, or fellow workers.
EQUIPMENT USED		**OPERATIONAL DESCRIPTION**

1. Office equipments	Examples are computer, typewriter, projector, and cassette player/recorder.
2. Hand Tools	(e.g., hammer, shovel, screwdriver)
3. Power Tools	(e.g., radial saw, reciprocating saw, drill, pheunomatic hammer)
4. Vehicles	(e.g., automobile, truck, tractor, lift)
HAZARDS	**OPERATIONAL DESCRIPTION**
1. Hazardous Situation	a. Proximity to moving, mechanical parts. b. Exposure to electrical shock. c. Working in high, exposed places. d. Exposure to radiant energy. e. Working with explosives & exposure to toxic or caustic chemicals.

Table 2. Job Description design guidelines

Though preparing job description and job specification are not legal requirements yet play a vital role in getting the desired outcome. These data sets help in determining the necessity, worth and scope of a specific job.

The main purpose of a Job Description is to collect job-related data in order to advertise for a particular job. It helps in attracting, targeting, recruiting and selecting the right candidate for the right job. It s done to determine what needs to be delivered in a particular job; it clarifies what workers are supposed to do if selected for that particular job opening. Also, it provides staff a clear view what kind of candidate is required by a particular department or division to perform a specific task or job.

Job Description and Job Specification are two integral parts of job analysis. They define a job fully and guide both employer and employee on how to go about the whole process of recruitment and selection.

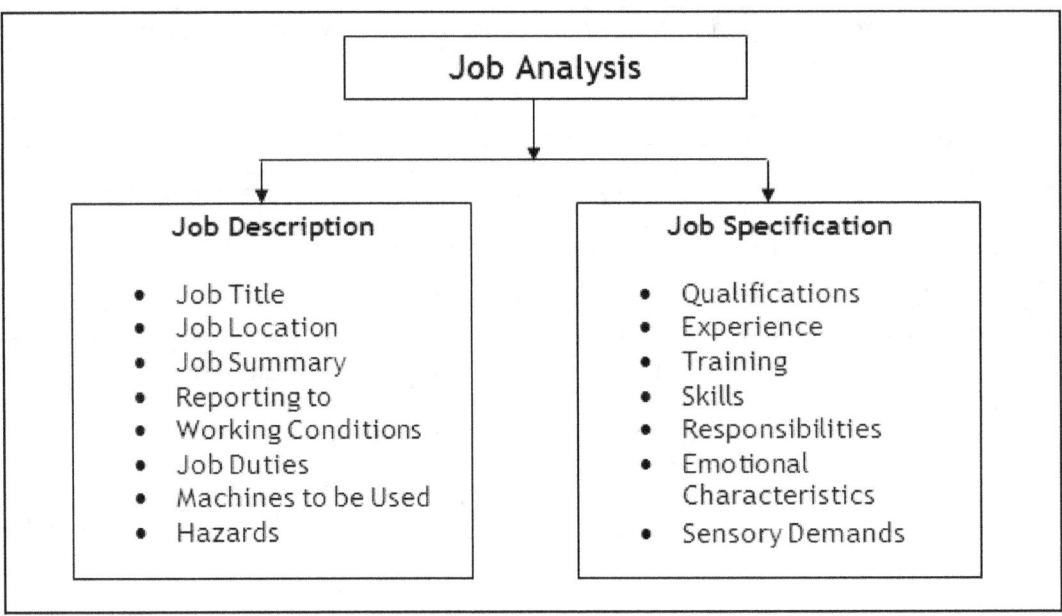

Figure 10. Job Analysis & its components

Both data sets are extremely relevant for creating a right fit between job and talent, evaluate performance and analyze training needs and measuring the worth of a particular job.

Moreover, Job Specification translates the job description into terms of the human qualifications, which are required for performance of a job. They are intended to serve as a guide in hiring and job evaluation.

Job Specification is a *written statement of qualifications, traits, physical and mental characteristics* that an individual must possess to perform the job duties and discharge responsibilities effectively – developed effectively with the co-operation of personnel department and various technical supervisors in the whole organization .

Job Specification Information

The first step in programming job specification is to prepare a list of all jobs in the company and where they are located. The second step is to secure and write up information about each of the jobs in a company that is usually information about each of the jobs in a company. Usually this information includes:

1. **Physical specifications:** - Physical specifications include the physical qualifications or physical capacities that vary from job to job. Physical qualifications or capacities
2. Include **Physical Features** like height, weight, chest, vision, hearing, ability to lift weight, ability to carry weight, health, age, capacity to use or operate machines, tools, equipment etc.
3. **Mental specifications:** - Mental specifications include ability to perform, arithmetical calculations, to interpret data, information blue prints, to read electrical circuits, ability to plan, reading abilities, scientific abilities, judgment, ability to concentrate, ability to handle variable factors, general intelligence, memory etc.
4. **Emotional and social specifications:** - Emotional and social specifications are more important for the post of managers, supervisors, foremen etc. These include emotional stability, flexibility, social adaptability in human relationships, personal appearance including dress, posture etc.
5. **Behavioral Specifications:** - Behavioral specifications play an important role in selecting the candidates for higher-level jobs in the organizational hierarchy. This specification seeks to describe the acts of managers rather than the traits that cause the acts. These specifications include judgments, research, creativity, teaching ability, maturity trial of conciliation, self-reliance, dominance etc.

REVIEW QUESTIONS:

1. Create a simple job description for the following engineering post:
   - *Cadet Civil Engineering* (for DSH Railway Construction)
   - *Junior Electrical Engineer* (for Toms Electric Cooperative)
   - *Engineering Manager* (knowledgeable in Six Sigma: for multinational company specializing in oil and crude industry)

2. Design a psychological test or special aptitude test for a Mechanical Engineering job description to determine manual dexterity, ingenuity & judgment.

3. Gather at least (15) fifteen engineering job post and develop its job description and requirements based on the current market and industrial demand.

RECOMMENDED READINGS

A. Books

Bartram, Dave, (2005). *The Great Eight Competencies: A Criterion-Centric Approach to Validation Journal of Applied Psychology* 2005, Vol. 90, No. 6, 1185–1203

Bemis, Stephen E., Holt-Belenky, Ann, Soder, Dee Ann (1983). *Job Analysis: an Effective Management Tool*, Washington, D.C., Bureau of National Affairs.

Cascio, Wayne F., Aguinis, Herman. (2011). *Applied Psychology in Human Resource Management, Seventh Edition*. Upper Saddle River New Jersey, Prentice Hall

Shippmann, Jeffery S., Ash, Ronald A, Battista, Mariangela, Carr, Linda, Eyde, Lorraine, D., Pearlman, Kennery, Prien, Erich P., Sanchez, Juan I. (2000) *The practice of competency modeling. Personnel Psychology*, Vol. 53, pp703-740

Sison, Perfecto. *Personnel and Human Resources Management*, Manila. Rex Publishing, 1982

B. E-books & Internet

Hurst, Carol L. "Position Analysis Questionnaire". Arizona State University ASU Office of Human Resources

"Job Analysis for changing workplace" available at orkspace.library.yorku.ca

"HR Guide to the Internet: Job Analysis" available at http://www.job-analysis.net/

"Job Description" available at http://www.psnacet.edu.in/

"Job Analysis & Description" available at https://www.aub.edu.lb/

"Importance of Job Analysis" available at http://www.educationobserver.com/

"Job Description & Job Specification" available at http://managementstudyguide.com

www.ingramcontent.com/pod-product-compliance
Lightning Source LLC
Chambersburg PA
CBHW080846170526
45158CB00009B/2647